Monique Alves F. de Moraes Freitas

Complete guide to evaluating measurement uncertainty

Monique Alves F. de Moraes Freitas

Complete guide to evaluating measurement uncertainty

Evaluation of measurement uncertainty with coordinate measuring machine, profile projector and tooling microscope

ScienciaScripts

Imprint

Any brand names and product names mentioned in this book are subject to trademark, brand or patent protection and are trademarks or registered trademarks of their respective holders. The use of brand names, product names, common names, trade names, product descriptions etc. even without a particular marking in this work is in no way to be construed to mean that such names may be regarded as unrestricted in respect of trademark and brand protection legislation and could thus be used by anyone.

Cover image: www.ingimage.com

This book is a translation from the original published under ISBN 978-620-4-19571-1.

Publisher:
Sciencia Scripts
is a trademark of
Dodo Books Indian Ocean Ltd. and OmniScriptum S.R.L publishing group

120 High Road, East Finchley, London, N2 9ED, United Kingdom
Str. Armeneasca 28/1, office 1, Chisinau MD-2012, Republic of Moldova, Europe
Printed at: see last page
ISBN: 978-620-7-98941-6

CONTENTS

CHAPTER I - INTRODUCTION

The industrial development of societies and, consequently, the fierce competition for the market have led to the realisation that obtaining competitive products depends on quality planning right from the product and process design stage. This means that the metrological quality and quality certification of manufactured products has become a major differentiator for competitive advantage in national and international markets.

Measuring therefore becomes an indispensable factor in winning over and satisfying a significant number of customers, since it makes it possible to quantify the determining quantities in the generation of a good or service.

However, there is currently an obligation to present a statement about the confidence associated with the result of every measurement, known as measurement uncertainty.

Uncertainty reflects doubt about the validity of a measurement result due to the lack of exact knowledge of the value of the measurand. Without an indication of uncertainty, measurement results cannot be compared, either with each other or with reference values given in manufacturer specifications, catalogues or technical standards. Nor can they be compared with previously published values in articles, dissertations and theses.

It is therefore necessary to have a readily implemented, easily understood and generally accepted procedure for characterising the quality of a measurement result, i.e. for evaluating and expressing its uncertainty.

The document "Guide to the expression of uncertainty in measurements" (popularly known as GUM), published in 1993, was created to standardise the procedures for evaluating and expressing measurement uncertainty. This guide presents general criteria and rules for expressing and combining the individual uncertainties that affect the inspection process and thus determining measurement uncertainty. Due to its extreme relevance, this document was translated in Brazil by a group of experts.

Understanding and applying the GUM (2003) is no easy task. Among other things, it requires extensive metrological and statistical knowledge, the manipulation of large amounts of data and mathematical calculations which, in certain situations, can be very complex. As such, assessing uncertainty can become tedious, time-consuming and is inevitably subject to calculation errors. An example of this is the mathematical model proposed by Vieira Sato (2003) for evaluating the measurement uncertainty of the circularity deviation in coordinate measuring machines, which requires the solution of approximately 50 partial derivatives.

For these reasons, many people refuse to apply the methodology proposed in GUM (2003), even though they are aware of the importance of evaluating measurement uncertainty.

Therefore, this work aims to implement electronic spreadsheets to evaluate the uncertainty of measurements made with a coordinate measuring machine, a profile projector and a tooling microscope, based on the uncertainty calculation methodology proposed in GUM (2003). They were prepared using the *Excel* application, which allows the uncertainty to be estimated simply and quickly.

The spreadsheets have been developed so that the user only has to enter the following data manually:

- resolution of the measurement system;

- number of decimal places in the resolution;

- number of readings or measurement cycles;

- values of the readings when the measurand is obtained directly from the indication of the measuring instrument or;

- values of the measurand when mathematical expressions are used to determine it from the indications of the measuring instrument;

- temperature values collected during the measurements;

- values of the geometric deviations relative to the components of the measurement system, if they are considered to be sources of uncertainty;

- uncertainty associated with the calibration of the measurement system;

- effective degrees of freedom of the calibration standard uncertainties considered.

The results are;

- standard uncertainty of each influence variable;

- combined standard uncertainty associated with the measurement of the measurand;

- effective degree of freedom of the measurement;

- coverage factor;

- expanded uncertainty associated with the measurement of the measurand;

- probability of coverage (95.45 %);

- sensitivity coefficients of the input quantities;

- estimation of the measurand.

Among the many positive aspects of the spreadsheets are the following: they are open, so users can modify and adapt them to new applications and measurement strategies; they allow quick analysis and identification of the influencing variables that most contribute to the final uncertainty, making it possible to improve measurement processes. In addition, the mathematical expressions used to calculate the different parameters are easily visualised by accessing the corresponding cell. Finally, all the information relating to the evaluation of measurement uncertainty is summarised in a table, compatible with Microsoft Word.

Therefore, it is known that measuring alone is not enough. It is now mandatory to present measurement results together with the associated uncertainty values so that they comply with current technical standards. This is because measurement results expressed only as the arithmetic mean are meaningless because they do not provide complete information about the measurement. Even if the standard deviation is stated alongside the mean, several influencing variables are not taken into account. These include the uncertainty related to the calibration of the measurement system, the temperature deviation from 20 °C, among others.

Therefore, the development of electronic spreadsheets for calculating measurement uncertainty is justified by the need to popularise the GUM (2003) and measurement uncertainty calculation scripts, so that all those who take measurements can assess uncertainty simply, safely and quickly, even if they are unfamiliar with the basics of calculation.

Estimating and declaring the uncertainty will increase the reliability of the measurements made, as it will allow measurement results to be presented in accordance with current standards. This is a necessary condition for increasing the scientific rigour of the work carried out and the publications resulting from it.

This book is based on the monograph defended by Moraes (2011).

CHAPTER II - LITERATURE REVIEW

The literature review begins with the study and presentation of two basic documents, ISO/IEC 17025 (2005) and the Guide to the Expression of Measurement Uncertainty (GUM, 2003). This is followed by published works on the subject in question.

2.1. NBR ISO/IEC 17025 (2005): General requirements for the competence of testing and calibration laboratories

The ISO/IEC 17025 (2005) standard contains all the specifications and requirements that laboratories responsible for carrying out tests and calibrations must fulfil in order to guarantee reliable results.

The adoption and implementation of this standard is essential to ensure the quality of all services provided by metrological laboratories, including research. It should be emphasised that compliance with this standard guarantees compliance with all the requirements already contained in NBR ISO 9001 (2000) and NBR ISO 9002 (2000).

NBR ISO/IEC 17025 (2005) is divided into five parts: objective, normative references, terms and definitions, management requirements and technical requirements. It discusses standardised methods, non-standardised methods and methods developed by the laboratories themselves, and is applicable to any entity that carries out tests and calibrations.

According to ISO/IEC 17025 (2005), the laboratory is responsible for all the activities it carries out and must therefore have qualified managerial and technical staff. It must also appoint a quality manager and draw up documents to inform about appropriate practices and procedures, including ensuring compliance with the standard.

All documents belonging to the laboratory, including standards, regulations, test methods or calibrations, drawings, specifications, *software* and manuals must be controlled by those responsible. They must be available for consultation by all staff as soon as they are approved.

The laboratory must have the equipment needed to carry out the activities that are part of its scope, as well as documents containing basic information about them. All instruments must be calibrated before use and, if any of them are damaged, they must be immediately marked and isolated. Only after calibration will they be allowed to be used again.

Calibration aims to ensure traceability to primary standards, both for measuring instruments and laboratory standards. In this sense, NBR ISO/IEC 17025 requires that measurement uncertainty be declared at all levels of the traceability chain.

After taking any measurement, the uncertainty must be calculated using appropriate methods. When it is not possible to use all the statistical criteria, at least a reasonable estimate of this parameter should be made in order to get an idea of the dispersion of the measured values.

There are several factors that influence the total measurement uncertainty and, depending on the type of test and/or calibration, they contribute with different intensities. Because of this, laboratories must take them into account in their technical testing or calibration procedures and in staff training.

Measurement uncertainty is defined as a non-negative parameter that characterises the dispersion of the values assigned to a measurand, based on the information used (VIM, 2009). Without its indication, a laboratory cannot compare its results with those of other laboratories or with those specified in catalogues, manuals and technical standards.

Measurement uncertainty exists because various factors affect the accuracy and precision of a measurement system. In order to issue technically valid measurement results, detailed studies are carried out of all the sources of uncertainty that could interfere with the value indicated by an instrument.

Sources of uncertainty include variations associated with the measuring instrument, the operator, the ambient conditions, the properties and conditions of the item being tested, among others.

Due to the need to evaluate and express measurement uncertainty in a uniform way throughout the world, after years of study the document "Guide to the expression of uncertainty in measurements" was published in 1993. And in 1997, due to its importance, this document was translated into Portuguese under the title "Guia para a expressão da incerteza de medição" (Guide to the expression of measurement uncertainty).

2.2. Guide to expressing measurement uncertainty (GUM)

Basically, the GUM (2003) establishes the general rules and criteria for expressing and combining the individual uncertainties that affect the measurement process and thus determining the total measurement uncertainty (Fig. 1), which can follow various levels of accuracy and complexity.

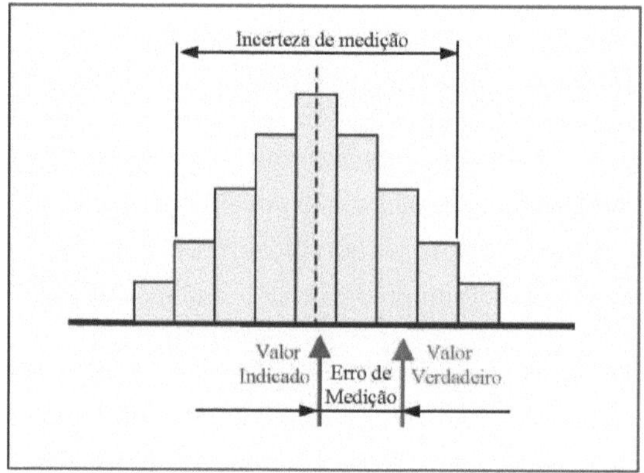

Figure 1: Measurement uncertainty

GUM (2003) conceptualises three types of uncertainty: standard uncertainty, combined standard uncertainty and expanded uncertainty. The first of these is related to each influence quantity and is obtained by analysing each variable individually. Once the effect of these quantities is known, it is possible to relate them using the Law of Propagation of Uncertainties, thus obtaining the combined standard uncertainty. In turn, the expanded uncertainty is the result of multiplying the value of the combined standard uncertainty by a factor, defined according to the desired level of coverage.

This guide proposes that, in the initial stage of calculating uncertainty, all the variables that influence the measurement result should be identified. The number and type of influencing quantities vary according to the measurement system and the type of measurand being analysed.

These uncertainty components are categorised into two groups: type A and type B assessments. Both are based on probability distributions and the uncertainty components resulting from each type are quantified by variances or standard deviations.

According to GUM (2003), type A standard uncertainty is that obtained from a statistical analysis of a series of observations of a measurand, assuming a normal distribution (Fig. 2) or any other distribution. An uncertainty component obtained from a Type A evaluation is characterised by a standard deviation that takes into account random fluctuations and influences considered to be constant in the results of a given experiment.

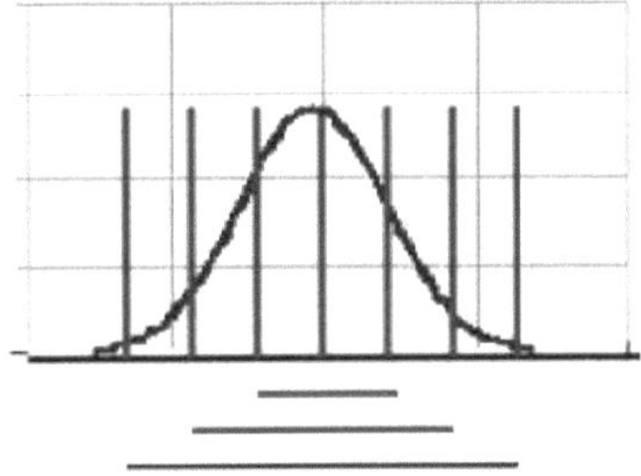

68.26%=> one standard deviation
95.45%=> two standard deviations
99.73% => three standard deviations

Figure 2: Normal distribution

Basically, type A standard uncertainty is obtained from a probability density function derived from the observation of a frequency distribution, i.e. based on a series of observations of the quantity. The set of readings of the indications of a measuring instrument is an example of a variable whose uncertainty is classified as type A, with a normal distribution and n-1 degrees of freedom. It can be calculated using Eq. (1).

(1)

$$u(x) = \sqrt{\frac{s^2}{n}}$$

Where "s" is the experimental standard deviation and "n" is the number of elements in the sample.

In contrast, type B standard uncertainty is obtained by other means, such as from manuals, manufacturers' specifications, calibration certificates or from previous experience. Depending on the amount of information available, it assumes one or other distribution.

The rectangular distribution (Fig. 3) is used when it is possible to estimate only the upper and lower limits for Xi and establish that the probability that the value Xi belongs to the interval (a-, a+) is one and the probability that the value Xi is outside this interval is zero. If there is no specific knowledge of possible values of Xi within the interval, it can be assumed that it is equally likely that Xi is anywhere in the interval, and consequently its degree of freedom is infinite (LINK, 1997).

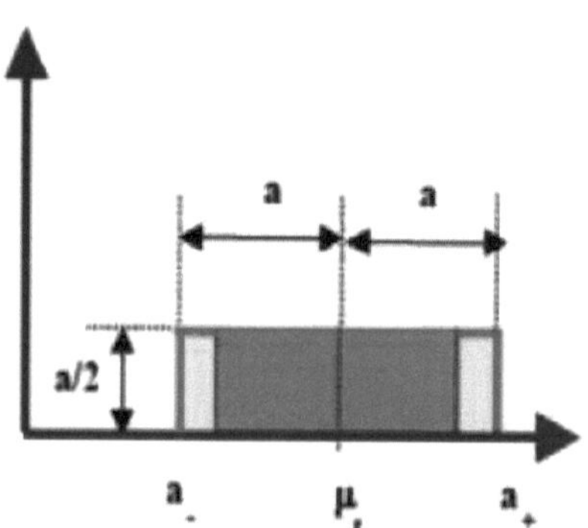

Figure 3: Rectangular distribution

In this case, the standard uncertainty of type B is estimated by Eq. (2).

$$u(x) = \frac{a}{\sqrt{3}}$$

(2)

Examples of sources of uncertainty with this type of distribution: temperature gradients, temperature deviation from 20 °C, measurement system resolution, hysteresis, among others.

If it is more likely to expect a value that is close to the midpoint than close to the limits of the interval, then the symmetrical trapezoidal distribution should be adopted (Fig. 4).

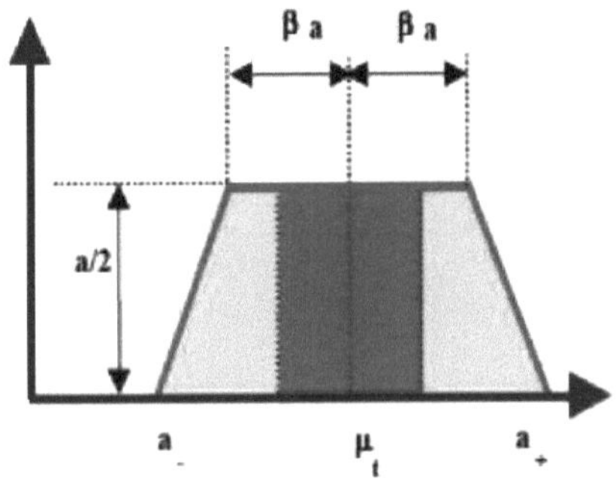

Figure 4: Trapezoidal distribution

The type B standard uncertainty of a variable with this distribution is given by Eq. (3).

(3)

$$u(x) = \frac{a.\sqrt{(1+\beta^2)}}{\sqrt{6}}$$

If there is more knowledge about the distribution of the possible values of the quantity, the probability distribution changes to a triangular distribution (Fig. 5), with infinite degrees of freedom, which can evolve into a normal distribution. Examples of sources of uncertainty with a triangular distribution are: deviation in the flatness of the measuring surfaces of an instrument, deviation in parallelism between the measuring surfaces, among others.

The type B standard uncertainty associated with a quantity with a triangular distribution is expressed by Eq. (4).

$$u(x) = \frac{a}{\sqrt{6}}$$

(4)

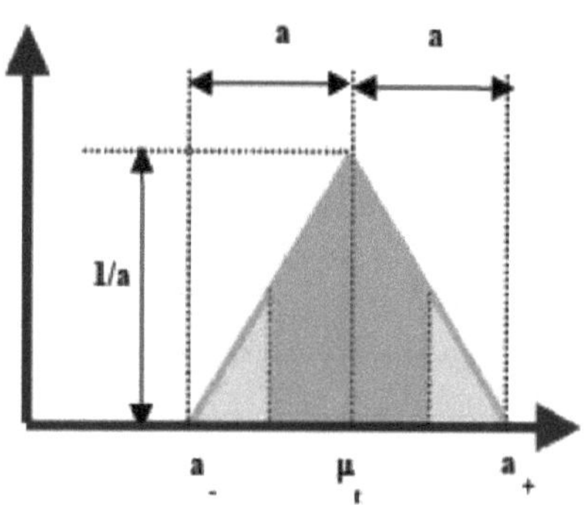

Figure 5: Triangular distribution

Once all the standard uncertainty values are known, the combined standard uncertainty is calculated. To do this, a mathematical model must be defined beforehand, as it is the basis for applying the Law of Propagation of Uncertainties. This model relates all the variables that influence the measurement process and is unique for each type of quantity measured. This function is given by Eq. (5).

$$Y = f(X_1, X_2, \ldots, X_N).$$

(5)

Taking y as the estimate for the response Y and x_1, x_2, ...x_n as the estimates for the input variables Xi (i= 1,...,N), the combined standard uncertainty, called u_c(y), is obtained from Eq. (6).

(6)

$$u_c^2(y) = \sum_{i=1}^{n} \left(\frac{\partial f}{\partial x_i} \right)^2 u^2(x_i) + 2 \sum_{i=1}^{n-1} \sum_{j=i+1}^{n} \frac{\partial f}{\partial x_i} \frac{\partial f}{\partial x_j} u(x_i) \cdot u(x_j) \cdot r(x_i, x_j)$$

Where u(x) is the standard uncertainty of the quantity x, $r(x_i, x_j)$ is the correlation coefficient between the estimates x_i and xj and the terms of the partial derivatives of the function in relation to each input variable indicate the sensitivity coefficients. The magnitude of these coefficients represents the contribution of each source of uncertainty to the value of the total uncertainty.

The second term in Eq. (6) will only exist when there is a correlation between the input quantities xi and xj , i.e. when $r(x_i, x_j)\ \phi\ 0$.

Although the combined standard uncertainty can be universally used to express the uncertainty of a measurement result, it only covers 68.27 % and is not suitable for most dimensional metrology applications. There is therefore a need to expand it in order to achieve greater coverage. In other words, the expanded uncertainty, U_p, must be estimated using Eq. (7).

$$U_p = k \cdot u_c$$

(7)

The coverage factor k in Eq (7) is chosen according to the confidence level specified for the interval, which is generally between 2 and 3 for a normal probability distribution. The coverage factor assumes 2 and 3, respectively, for an interval with a confidence level of 95.45 % and 99.73 %.

When the number of readings is small, characterising a small sample, this approximation for the coverage factor is not convenient. In this case, the central value theorem should be used together with the t-student table to provide a value for k based on the effective degree of freedom of the standard measurement uncertainty and the confidence level adopted.

The calculation of the effective degree of freedom is based on the Welch-Satterwaite equation, as expressed in Eq. (8).

$$v_{ef(WS)} = \frac{u_c^4(y)}{\sum_{i=1}^{n} \frac{u_i^4(x)}{v_i}}$$

(8)

The measurement result, taking into account the combined standard uncertainty, is given by $y \pm u_c$, and, in relation to the expanded uncertainty, it is expressed by $y \pm u_p$, where y is the estimate of Y, being its average value.

According to GUM (2003), the steps for determining the uncertainty of a measurement result are basically as follows:

- identify the influencing variables;

- express the measurand as a function of all the quantities that affect the measurement process, using a mathematical model;

- determine, statistically or by other means, the components of the measurement uncertainty;

- evaluate the standard uncertainty of each input estimate, classifying it as type A or type B;

- evaluate the covariances if the input quantities are correlated;

- calculate the measurement result, i.e. the estimate y of the measurand Y, which represents the average value of the measurements;

- determine the combined standard uncertainty of the measurement result by applying the Law of Propagation of Uncertainty;

- calculate the expanded uncertainty;

- report the measurement result together with the expanded uncertainty, the coverage factor and the corresponding coverage probability.

Finally, it is recommended that all the information regarding the evaluation of measurement uncertainty be summarised in a table.

2.2. Work on uncertainty and spreadsheets

The most commonly used measuring systems in the metalworking industry are: universal callipers, external micrometers, dial gauges, tape measures, scales, height gauges, microscopes and coordinate measuring machines. Currently, the metrological reliability of measurements with these measuring systems is sought mainly through their calibration. However, guaranteeing this reliability includes other aspects, ranging from selecting the most suitable measuring system to correctly expressing the result together with its uncertainty.

The estimation of measurement uncertainty generally requires the manipulation of large amounts of data, so the use of spreadsheets and specially developed computer programmes is recommended. Electronic spreadsheets are the most widespread among metrological laboratories, as they allow the uncertainty calculation methodology to be implemented using programmes found on the market, which are inexpensive to purchase and can also be used for other purposes.

One of the positive aspects of electronic spreadsheets is that they are practical for recording measurements, as well as allowing the results to be visualised quickly. In this way, a critical analysis of measurement uncertainty can be carried out. It is quick and easy to identify which uncertainty component contributes most to the final uncertainty, simply by checking which of the uncertainty contributions has the highest value. The comparison is easy and straightforward because all the uncertainty contributions are in the same unit of measurement as the measurand. It is worth noting that when the standard uncertainties do not have the same unit as the measurand, the respective sensitivity coefficients, which are not unitary, will allow them to have the same unit, so that they can then be combined with the others to calculate the combined standard uncertainty.

According to Kacker et al. (2007), the use of electronic spreadsheets provides transparency in the expression of measurement uncertainty and is also fundamental in inter-laboratory comparisons and in the evaluation of metrological laboratories.

Although they are important, GUM does not provide any guidance on the layout of these spreadsheets, so there is no standard formatting for them.

The use of spreadsheets can be greatly improved when they are used in conjunction with electronic document management (EDM) software. EDM tools provide greater reliability and accessibility when handling computer files, allowing spreadsheets to be opened for consultation or modification according to the permissions established for each user.

Before implementing the spreadsheets, it is necessary to carry out a study of the measurement systems, measurement uncertainty and the standardised methodology described in the GUM.

In the literature you can find various works on calculating measurement uncertainty, among them:

Franco (1996) presented the requirements for applying and assessing metrological reliability in metrology laboratories. These requirements make it possible to assess whether the laboratory is eligible to belong to the Brazilian Calibration Network (RBC). To this end, the measurement uncertainty of standard blocks was determined, and the results obtained were compared with those provided by the INMETRO recommendation. It was concluded that the initial proposal of the work was achieved, since it was possible to assess the efficiency of the laboratory to produce reliable results during the calibration of standard blocks.

Barp (2000) proposed procedures for guaranteeing metrological reliability in temperature measurement processes. An environment for analysing uncertainty, following the requirements of the GUM, was developed using well-established programmes. The Excel programme (Microsoft Co.) was used to help calculate the uncertainty and the Pspice programme (MicroSim Co.) was used to quantify the individual sources of uncertainty. A case study of a temperature measurement system was carried out, using the uncertainty management procedure and the proposed environment to help analyse uncertainty. The strong applicability of this method for evaluating the uncertainty of automated measurement systems was observed, bypassing many of the problems that discourage the use of GUM in such complex applications.

The author concluded in this work that the integrated system to aid uncertainty assessment proved to be viable and operational. Although using the system requires knowledge of instrumentation and metrology, it does not require very specialised knowledge in the various areas, making it accessible to professionals at various levels found in industry. Furthermore, the automation of the quantification of individual uncertainties and the calculation of the expanded uncertainty, through the spreadsheets implemented in Excel, allows for a considerable reduction in evaluation time.

Sallum, A.T., (2002) developed a model of an automated system based on modern software engineering techniques using the Unified Modelling Language, employing the philosophy of object-oriented programming. This made it possible to generate procedures to help the user reduce systematic effects; configure instruments; acquire measurement data; reject discrepant values; process data; calculate uncertainty and generate documents such as uncertainty spreadsheets and calibration certificates. The result of this work is a comprehensive computer programme, which enables the generation of consistent procedures for automating the calibration of various instruments and also of different calibration methods for the same measurement system.

Grachanen (2002) has developed software for calculating uncertainty that allows measurement uncertainty to be calculated analytically by implementing the mathematical models that describe the measurement system in use. This programme arose from the need to document measurements and the results obtained, together with the associated uncertainty, in a simple, straightforward way that complies with the GUM. This software has made a major contribution to the global measurement community that has had access to it.

Stempniak (2004) developed a study based on the methodology of the Guide to the Expression of Measurement Uncertainty, Monte Carlo Simulation (MCS) and the work of the Software Support for Metroogy (SSfM) programme. As a result, it was possible to describe the rationale, architecture, implementation and validation of a software tool called AUTOLAB INTUIT, which is capable of modelling, automating and evaluating measurement systems and is not limited to specific areas of metrology. The tool was designed and applied in various practical cases, carrying out automated measurement acquisition, calculating uncertainties and issuing reports in line with international standards. It differs from other similar tools in particular due to its more intuitive graphical interface and its use of more modern and robust methodologies.

Matusevich (2007) implemented a tool for calculating parameters and uncertainties in tensile testing. This arose from the requirement for testing and calibration laboratories to have and apply procedures for assessing uncertainty. Although there are no specific guidelines for evaluating the uncertainties associated with the values measured in a tensile test, the recommended practice is to use the method standardised by the GUM. The software developed by the author was called "IncerTI", which allows the parameters of a tensile test and the associated uncertainties to be calculated. This programme has been used in the INTI-Córdoba Testing Laboratory since 2007 and, due to its continuous use, has been improved, resulting in increasingly robust, easy-to-use software with greater capacity.

Lemos *et al.* (2009) presented four mathematical models for calculating the uncertainty of different measurements made with universal callipers. These models are based on the GUM methodology. This work emphasised that in order to estimate measurement uncertainty properly, it is essential to know the constructive characteristics of the measuring system and its operating principle.

Oliveira and Mesquita (2009) presented the development and operation of the measurement module of a computer programme for metrological control and definition of the conformity zone for a given product.

This module was developed in Excel, using routines prepared in VBA *(Visual Basic for Application)*. Initially, the measurement system was selected and then the data was entered manually, including: the resolution of the system, the coverage factor for the expanded uncertainty and the readings taken. Once the measurand has been measured, this data must be entered into the spreadsheets for subsequent analysis. Using the spreadsheets developed, the value of k (coverage factor) is obtained directly from Excel using the INVT function, which returns the value of the Student's t distribution as a function of probability and effective degrees of freedom. It is important to note that the programme can generate the measurement uncertainty for three different reliability levels: 90; 95.45 and 99.73

%, which can be selected via a checkbox. Thus, using the computer programme presented in this work, it is possible to manage the measurement data in order to express its result reliably, as well as defining an acceptance and rejection zone for parts based on the measured values.

By determining the zone of conformity via a computer programme, i.e. without the need for manual calculations, there is a gain in terms of the reliability of the manufactured product, since it is still very common in industrial practice to adopt the specification zone as the range for guaranteeing conformity.

Moraes *et al.* (2010) presented a methodology for calculating the uncertainty associated with the calibration of analogue dial gauges using a Universal Measuring Machine. This calibration, however, is not an easy task, requiring highly accurate equipment and a highly trained operator. In addition, when calculating the uncertainty associated with calibration, it was realised that several influencing variables had to be taken into account and consequently a large amount of data had to be manipulated, which led to the need to implement a spreadsheet for this purpose.

Jornada (2009) implemented a guideline for evaluators of laboratories in the RS Metrological Network, which considers measurement uncertainty as one of the requirements to be taken into account when evaluating them. The author drew up a flowchart explaining the steps for expressing measurement uncertainty, as well as listing the typical uncertainty components for testing and calibration areas. He also drew up a checklist to help evaluators when assessing uncertainty. As a result, the evaluators' level of knowledge regarding uncertainty improved.

According to the author, when applying the GUM method, it is advisable to use spreadsheets to gather all the data on measurement uncertainty. The uncertainty spreadsheet lists information such as uncertainty components and their respective estimates, sensitivity coefficients, uncertainty contributions, combined standard uncertainty, the coverage factor k and expanded uncertainty. In this context, the uncertainty guideline emphasises that validation by manually reproducing the calculations performed on the electronic uncertainty spreadsheets used by the laboratory is only one way of guaranteeing this. Another possibility is to check the formulae in these spreadsheets. In this case, it should be checked that all the formulae in the spreadsheet are correct and that they reference the relevant cells in the spreadsheet. Finally, the guide emphasises that the assessor should look for laboratory records that show this assessment and that there should be a periodic validation of the spreadsheets by the laboratory, in order to guarantee the correct calculation of uncertainty.

CHAPTER III - METHODOLOGY

The following steps were proposed for the development of this work:

I) *Identification of measurement systems and measurands.* In this stage, the measurement systems of interest and the measurands commonly measured with them were identified.

II) *A thorough study of measurement systems in order to correctly identify all sources of uncertainty.* To this end, the operating principle, constructive characteristics and sources of error relating to the measurement systems were studied. Next, the measurands (output variables) and the variables of influence (input variables) were identified.

III) *Mathematical modelling of the measurement process.* The output variable was written as a function of all the input variables. This model represents the starting point for the application of GUM.

IV) *Preparation of calculation scripts.* A script was drawn up to calculate the standard, combined standard and expanded uncertainties, depending on the information available.

V) *Implementation of the spreadsheets.* In this stage, the uncertainty calculation scripts were implemented in spreadsheets using *Excel* tools. A spreadsheet was created for each of the measurands. At the same time, checks were made to ensure that the equations entered were correct and that the cells referenced the necessary terms.

VI) *Carrying out experimental tests.* Experimental tests were planned and carried out to collect the data needed for the simulation and also to validate the spreadsheets.

VII) *Validation of the spreadsheets.* In this stage, the uncertainties were calculated using the experimental data and the spreadsheets were validated.

3.1. Identification of measurement systems and measurands

Initially, the measurement systems of interest and the measurands to be measured with these measurement systems were identified and the spreadsheets implemented (Table 1).

Table 1. Measurement systems and measurands considered

Measuring systems	Measuring
Coordinate measuring machine	Circularity deviation Cylindricity deviation Straightness deviation Linear dimension Angle
Profile projector	Linear dimension Angle
Tooling microscope	Linear dimension Angle

3.2. Study and modelling of measurement processes

The particularities of measurement systems, mathematical models for evaluating uncertainty and calculation scripts are presented below.

3.2.1. Measurement uncertainty with a coordinate measuring machine

Coordinate measuring machines (CMMs) are the metrological instruments responsible for meeting the needs of modern industry.

They are characterised by their simplicity of operation, flexibility and the fact that they allow complex structures to be measured extremely quickly and accurately, as well as the simultaneous control of various metrological characteristics of a part (Kunzmann, 1988). It can be said that these machines have revolutionised dimensional metrology.

Based on its operating principle, it is possible to know the coordinates (X, Y and Z) of the points belonging to the machine's working volume. From these coordinates, using the least squares method, the different geometries can be calculated (circle, sphere, cone, cylinder, plane, etc.).

As the sources of uncertainty using the MMC are similar for the different types of measurement, a common procedure will be presented for all measurands.

It should be noted that the uncertainty associated with the calibration of the CMM is expressed in the X, Y and Z axis directions. The uncertainty to be considered in the model depends on the characteristic measured.

For CMM measurements, the sources of uncertainty are: variability of the readings; digital resolution of the CMM; standard uncertainty associated with the calibration of the CMM, in the directions considered during the measurement; CMM probing error; temperature variation during measurements and temperature deviation from 20 °C.

The proposed mathematical model is expressed according to Eq. (9) (SOUZA et al., 2011).

$$C = s(L_{MMC}) + \Delta R_{MMC} + \Delta I_{CMMC} + \Delta E_A + L_0.\Delta T.(\alpha_{Pe} + \alpha_{MMC}) + L_0.\delta T.(\alpha_{Pe} + \alpha_{MMC}) \tag{9}$$

In which:

$s(L_{MMC})$: standard deviation of the measurand values;

ΔR_{MMC} : correction due to the MMC resolution;

ΔI_{CMMC} correction associated with the standard uncertainty of the CMM calibration, in the directions considered during the measurement;

ΔE_A: correction due to MMC probing error;

δT: correction associated with temperature variation during measurements;

ΔT correction associated with the deviation of the ambient temperature from 20 °C;

α_{MMc}: coefficient of linear thermal expansion of the MMC scales;

α_{Pe} coefficient of linear thermal expansion of the workpiece material;

L_0 : value of the measurand.

The standard uncertainties associated with MMC calibration are expressed by Eqs.

(10) a (12). This uncertainty depends on the axes that are moved during the

measurement of the characteristic in question.

1. For linear measurements in the direction of a given axis, the

uncertainty associated with calibration in this axis. For the direction of the X, Y and Z axes we have, respectively:

$$I_{CMMC(x)} = \frac{U_{Px}}{k} \tag{10}$$

$$I_{CMMC(y)} = \frac{U_{Py}}{k} \tag{11}$$

$$I_{CMMC(z)} = \frac{U_{Pz}}{k} \tag{12}$$

2. If the measurement is made in the plane, then Eqs. (13) to (15).

$$I_{CMMC(x,y)} = \sqrt{I_{CMMC(x)}^2 + I_{CMMC(y)}^2} \tag{13}$$

$$I_{CMMC(x,z)} = \sqrt{I_{CMMC(x)}^2 + I_{CMMC(z)}^2} \tag{14}$$

$$I_{CMMC(y,z)} = \sqrt{I_{CMMC(y)}^2 + I_{CMMC(z)}^2} \tag{15}$$

3. For volumetric measurements, the uncertainty associated with the calibration of the machine takes into account the uncertainty declared for the three coordinate axes, according to Eq. (16).

$$I_{CMMC(x,y,z)} = \sqrt{I_{CMMC(x)}^2 + I_{CMMC(y)}^2 + I_{CMMC(z)}^2} \tag{16}$$

The uncertainty related to the probing error is given by Eq. (17), considering a triangular probability distribution.

$$u(\Delta E_A) = \frac{E_A}{\sqrt{6}} \tag{17}$$

Applying the Law of Propagation of Uncertainties to Eq. (9), we obtain Eq. (18).

$$u(C)^2 = \left(\frac{\partial C}{\partial s(L_{MMC})}\right)^2 \cdot (u_{s(L_{MMC})})^2 + \left(\frac{\partial C}{\partial \Delta R_{MMC}}\right)^2 \cdot (u_{\Delta R_{MMC}})^2 + \left(\frac{\partial C}{\partial \Delta E_A}\right)^2 \cdot (u_{\Delta E_A})^2 + \left(\frac{\partial C}{\partial \alpha_{Pe}}\right)^2 \cdot (u_{\alpha_{Pe}})^2 + \\ + \left(\frac{\partial C}{\partial \alpha_{MMC}}\right)^2 \cdot (u_{\alpha_{MMC}})^2 + \left(\frac{\partial C}{\partial \Delta T}\right)^2 \cdot (u_{\Delta T})^2 + \left(\frac{\partial C}{\partial \delta T}\right)^2 \cdot (u_{\delta T})^2 + \left(\frac{\partial C}{\partial \Delta I_{CMMC}}\right)^2 \cdot (\Delta I_{CMMC})^2$$

(18)

3.2.2. Measurement uncertainty with a profile projector

Basically, the profile projector is used to check small parts, especially those with a complex shape, by projecting the enlarged image of the part onto its glass screen. This screen has two perpendicular lines engraved on it, which can be used as a reference for linear and angular measurements. The projector also has a mobile coordinate table with two micrometric heads, or two linear scales, positioned at 90° and two scales for measuring angles,

For measurements of linear or angular dimensions with a profile projector, the sources of uncertainty in the measurement result are: variability of the readings; resolution of the projector; standard uncertainty associated with calibration of the projector; temperature variation during measurements and temperature deviation from 20 °C.

The proposed mathematical model is expressed according to Eq. (19).

$$C = s(L_{Pr}) + \Delta R_{Pr} + \Delta I_{Pr} + L_0.\Delta T.(\alpha_{Pe} + \alpha_{Pr}) + L_0.\delta T.(\alpha_{Pe} + \alpha_{Pr})$$

(19)

In which:

$s(L_{pr})$: standard deviation of the measurand values indicated by the projector;

ΔR_{pr} : correction due to projector resolution;

ΔI_{pr} correction associated with the standard uncertainty of the projector calibration;

δT: correction associated with temperature variation during measurements;

ΔT correction associated with the deviation of the ambient temperature from the reference temperature (20 °C);

α_{Pr} coefficient of linear thermal expansion of the material of the micrometre heads (for linear dimensions) or of the knurling (for angular dimensions) of the projector;

α_{Pr} coefficient of linear thermal expansion of the workpiece material;

L_0 : value of the measurand.

Applying the Law of Propagation of Uncertainties to Eq. (19), we obtain Eq. (20).

$$u(C)^2 = \left(\frac{\partial C}{\partial s(L_{Pr})}\right)^2 \cdot (u_{s(L_{Pr})})^2 + \left(\frac{\partial C}{\partial \Delta R_{Pr}}\right)^2 \cdot (u_{\Delta R_{Pr}})^2 + \left(\frac{\partial C}{\partial \alpha_{Pe}}\right)^2 \cdot (u_{\alpha_{Pe}})^2 + \left(\frac{\partial C}{\partial \alpha_{Pr}}\right)^2 \cdot (u_{\alpha_{Pr}})^2 +$$
$$+ \left(\frac{\partial C}{\partial \Delta T}\right)^2 \cdot (u_{\Delta T})^2 + \left(\frac{\partial C}{\partial \delta T}\right)^2 \cdot (u_{\delta T})^2 + \left(\frac{\partial C}{\partial \Delta I_{Pr}}\right)^2 \cdot (\Delta I_{Pr})^2 \tag{20}$$

3.3.1. Measurement uncertainty with a tooling microscope

With its robust construction and easy handling, the toolmaker's microscope is particularly suitable for the shop floor.

Applications for this measuring system include: measuring the pitch of threads by placing the workpiece between the tips; checking a profile using the revolver eyepiece for threads and the workpiece on V-shaped or prismatic supports; measuring the internal diameter; measuring the distance between holes using the double image eyepiece, which is also used for measuring grooves and angles.

For measurements of linear or angular dimensions with a tooling microscope, the sources of uncertainty in the measurement result are: variability of the readings; resolution of the microscope; standard uncertainty associated with the calibration of the microscope; standard uncertainty associated with the calibration of the microscope's optical system; temperature variation during measurements and temperature deviation from 20 °C.

The proposed mathematical model is expressed according to Eq. (21).

$$C = s(L_{Mi}) + \Delta R_{Mi} + \Delta I_{CMi} + \Delta I_{CSO} + L_0 . \Delta T . (\alpha_{Pe} + \alpha_{Mi}) + L_0 . \delta T . (\alpha_{Pe} + \alpha_{Mi}) \tag{21}$$

In which:

$s(L_{Mi})$: standard deviation of the measurand values indicated by the microscope;

ΔR_{Mi} correction associated with the resolution of the microscope;

ΔI_{CMi} correction due to the standard uncertainty of the microscope calibration;

ΔI_{CSO} : correction due to the standard uncertainty of the microscope's optical system calibration;

δT : correction associated with temperature variation during measurements;

ΔT: correction associated with the deviation of the ambient temperature from 20 °C;

α_{Mi} coefficient of linear thermal expansion of the material of the microscope scales;

α_{Pe} coefficient of linear thermal expansion of the workpiece material;

L_0 : value of the measurand.

Applying the Law of Propagation of Uncertainties gives Eq. (22).

$$u(C)^2 = \left(\frac{\partial C}{\partial s(L_{Mi})}\right)^2 .(u_{s(L_{Mi})})^2 + \left(\frac{\partial C}{\partial \Delta R_{Mi}}\right)^2 .(u_{\Delta R_{Mi}})^2 + \left(\frac{\partial C}{\partial \alpha_{Pe}}\right)^2 .(u_{\alpha_{Pe}})^2 +$$
$$+ \left(\frac{\partial C}{\partial \alpha_{Mi}}\right)^2 .(u_{\alpha_{Mi}})^2 + \left(\frac{\partial C}{\partial \Delta T}\right)^2 .(u_{\Delta T})^2 + \left(\frac{\partial C}{\partial \delta T}\right)^2 .(u_{\delta T})^2 + \tag{22}$$
$$+ \left(\frac{\partial C}{\partial \Delta I_{CMi}}\right)^2 .(\Delta I_{CMi})^2 + \left(\frac{\partial C}{\partial \Delta I_{CSO}}\right)^2 .(\Delta I_{CSO})^2$$

3.3. Implementation of spreadsheets

Once the mathematical models and measurement uncertainty calculation scripts had been defined, the spreadsheets were implemented. The *Excel* application, version 2007, was used for this purpose.

Implementation consisted of the following steps:

3.3.1. Defining the layout of spreadsheets

The spreadsheets have been implemented on a single screen, so that the user has access to all the information, both incoming and outgoing.

In addition, the spreadsheets are open, allowing the user to modify and adapt them to other measurement procedures for the measurand in question.

The information was divided into eleven groups, thus considering 11 fields, described below.

Field 1. Input data

In this field the user must manually enter the input data, such as:

- resolution value of the measurement system used;

- number of decimal places in the resolution;

- unit of the measurement system indication according to SI;

- number of readings or measurement cycles;

- reading values (when the value of the measurand corresponds to the value indicated by the measuring system) or

- calculated values for each cycle (when the measurand is obtained by means of a mathematical equation, from the indications of the measuring system);

- number and values of the temperature readings collected during the measurements;

- standard uncertainty associated with the calibration of the measurement system;

- values of the geometric deviations relative to the measurement system and which contribute to the final uncertainty.

The spreadsheet automatically provides the sample arithmetic mean and standard deviation values.

It should be noted that all the values provided by the spreadsheets are presented with a higher number of digits than the resolution, in order to reduce calculation errors.

Field 2. Calculation of standard uncertainty

In the second field, the equations that make it possible to calculate the standard uncertainty associated with each input variable were entered, as described in the calculation script. The spreadsheet then provides these values automatically.

Field 3. Calculation of sensitivity coefficients

In this field, the sensitivity coefficients of the input variables were calculated for cases where they were different from one.

Field 4. Degrees of freedom

In this field, *Excel* calculates the degrees of freedom associated with the variability of the readings or values of the measurand, as a function of the number of readings or measurement cycles. It also calculates the effective degrees of freedom of the magnitude of the temperature deviation from 20 °C.

The degrees of freedom for the other input variables are already in the spreadsheets or must be supplied by the user if there are incomplete cells, and they depend on the effective degrees of freedom of the calibration standard uncertainty, both of the measurement system in question and of the temperature sensor used.

Field 5. Calculation of the combined standard uncertainty

Using the uncertainty propagation law equation, *Excel* calculates the value of the combined standard uncertainty and provides its value in the corresponding field.

Field 6. Calculation of effective degree of freedom

Using the Welch-Satterwaite equation, previously entered into the spreadsheet, the value of the effective degree of freedom is calculated and displayed.

Field 7. Determining the coverage factor

The value of the coverage factor (k) is determined automatically using the "INVT" function, which depends on the level of coverage considered, which in this case was 95.45 %, and the value of the effective degree of freedom, which has already been calculated. There is therefore no need to use the t-student table.

Field 8. Calculation of expanded uncertainty.

In this field, the values of the combined standard uncertainty and the coverage factor are multiplied to obtain the expanded uncertainty.

Field 9. Rounding of the expanded uncertainty value according to the decimal places of the resolution.

As all the values calculated using the spreadsheet have many significant figures, the value of the expanded uncertainty should be rounded up to the number of digits in the resolution. This is necessary because this value will be declared together with the arithmetic mean of the measurand values.

Field 10 Expression of the measurement result.

The spreadsheet also shows how the measurement result should be declared so that it complies with current standards. In this way, the values of the arithmetic mean, the expanded uncertainty, the coverage factor and the corresponding coverage probability, usually 95.45 %, must be presented.

Field 11 - Table containing all the uncertainty assessment information

Finally, Field 11 shows, in a table, all the information regarding the calculation of measurement uncertainty, as recommended by the GUM (2003). This table is usually presented in calibration certificates and test reports.

This table is filled in automatically as the information is entered into the spreadsheet. It is compatible with Microsoft Word and can be transferred using the copy and paste options.

3.3.2. Carrying out simulations

Once all the calculation scripts had been implemented, simulations were carried out in order to identify possible errors. For this purpose, arbitrary values were entered.

CHAPTER IV - EXPERIMENTAL TESTS, RESULTS AND DISCUSSIONS

Experimental tests were planned and carried out in order to run simulations on the spreadsheets and then validate them. To this end, various measurands were measured according to the proposed calculation scripts.

The measurements were carried out at the Federal University of Uberlândia's Dimensional Metrology Laboratory (LMD) at a temperature of (20 ± 1) °C. During the measurements, the ambient temperature was monitored using a digital thermo-hygrometer with a resolution of 0.1 °C and a nominal range of - 20 to 60 °C.

4.1. Measuring with a coordinate measuring machine

The coordinate measuring machine used to take the measurements is from the manufacturer Mitutoyo and has the following characteristics: it is a movable bridge type; it has a single tip with a ruby ball with a diameter of 2 mm; a resolution of 1 pm; a measuring capacity or working volume of 400 x 400 x 300 mm for the X, Y and Z axes respectively.

The measurements taken with this measuring system were in relation to the characteristics of the part, as shown in Fig. 6.

Figure 6. Part to be measured with MMC

The diameter and roundness deviation of the part were measured by probing 9 points. The measurement values collected are shown in Table 2.

Table 2. Measuring the diameter and circularity deviation of the part using the MMC

Measurements	Measuring	
	Diameter (mm)	Roundness deviation (mm)
1	30,016	0,028
2	30,019	0,030
3	30,018	0,027
Average (mm)	30,018	0,028
Standard deviation (mm)	0,002	0,002

Tables 3 and 4 show the components of measurement uncertainty for these measurands.

Table 3. Calculation of diameter uncertainty using MMC

Uncertainty components						
Greatness	Estimate	IT	DP	GL	CS	Standard uncertainty
LMMC	30,018	A	Normal	2	1	0,00088
RMMC	0,001	B	Rectangular	O	1	0,00029
EA	0,0029	B	Triangular	O	1	0,00118
ICMMC	0,00122	B	Normal	2E- 12	1	0,00122
Combined standard uncertainty (u_c) in mm						0,00194
Effective degree of freedom veff						1,4128E-11
Scope factor k						12,706
Expanded uncertainty in mm						0,025

It can be seen that the result of measuring the diameter of the part is (30.018 ± 0.025) mm, for k = 12.71 and a probability of coverage of 95.45 %. In turn, the result of the circularity deviation measurement is (0.028 ± 0.005) mm, for k = 2.08 and 95.45 % coverage.

Table 4. Calculation of circularity deviation uncertainty using MMC

Uncertainty components						
Greatness	Estimate	TI	DP	GL	CS	Standard uncertainty
LMMC	0,028	A	Normal	2	1	0,00088
RMMC	0,001	B	Rectangular	O	1	0,00029
EA	0,0029	B	Triangular	O	1	0,00118
ICMMC	0,00163	B	Normal	7,9402	1	0,00163
Combined standard uncertainty (u_c) in mm						0,00222
Effective degree of freedom veff						20,334
Scope factor k						2,086
Expanded uncertainty in mm						0,005

The flatness deviation of the part was measured by probing 7 points.

The measurement values collected are shown in Table 5.

Table 5: Flatness deviation values of the part using the MMC

Measurements	Flatness deviation (mm)
1	0,033
2	0,030
3	0,035
Average (mm)	0,033
Standard deviation (mm)	0,003

Table 6 gives the components of the measurement uncertainty for this measurand.

Table 6. Evaluation of flatness deviation uncertainty using MMC

Uncertainty components						
Greatness	Estimate	TI	DP	GL	CS	Standard uncertainty
LMMC	0,033	A	Normal	2	1	0,00145
RMMC	0,001	B	Rectangular	O	1	0,00029
EA	0,0029	B	Triangular	O	1	0,00118
ICMMC	0,00163	B	Normal	7,9402	1	0,00163
Combined standard uncertainty (u_c) in mm						0,00250
Effective degree of freedom veff						12,53345
Scope factor k						2,179
Expanded uncertainty in mm						0,005

The height of the piece was measured by probing 7 points. It should be noted that this measurement was taken as the distance between a point on the stretcher and a plane. The measurement values collected are shown in Table 7.

Table 7. Measuring part height using MMC

Measurements	Height (mm)
1	60,224
2	60,200
3	60,231
Average (mm)	60,218
Standard deviation (mm)	0,016

Table 8 gives the components of the measurement uncertainty for this measurand. It can be seen that the measurement result is (60.218 ± 0.121) mm, for k = 12.71 and 95.45 % coverage.

Table 8. Evaluation of part height measurement uncertainty using MMC

Uncertainty components						
Greatness	Estimate	IT	DP	GL	CS	Standard uncertainty
LMMC	60,218	A	Normal	2	1	0,00939
RMMC	0,001	B	Rectangular	w	1	0,00029
EA	0,0029	B	Triangular	w	1	0,00118
ICMMC	0,00105	B	Normal	1E- 12	1	0,00105
Combined standard uncertainty (uc) in mm						0,00952
Effective degree of freedom veff						8,22484E-09
Scope factor k						12,706
Expanded uncertainty in mm						0,121

The part's straightness deviation was measured by probing 5 points.

The measurement values collected are shown in Table 9.

Table 9. Measuring part straightness deviation using MMC

Measurements	Deviation Straightness (mm)
1	0,034
2	0,025
3	0,025
Average (mm)	0,028
Standard deviation (mm)	0,005

Table 10 gives the components of the measurement uncertainty for this measurand. It can be seen that the measurement result is (0.028 ± 0.043) mm, for k = 12.71 and 95.45 % coverage.

Table 10. Evaluation of the uncertainty of the part's straightness deviation using the MMC

Uncertainty components						
Greatness	Estimate	IT	DP	GL	CS	Standard uncertainty
LMMC	0,028	A	Normal	2	1	0,003
RMMC	0,001	B	Rectangular	w	1	0,00029
EA	0,0029	B	Triangular	w	1	0,00118
I_{CMMC}	0,0011	B	Normal	1E- 12	1	0,00110
Combined standard uncertainty (u_c) in mm						0,00342
Effective degree of freedom veff						1,36774E-10
Scope factor k						12,706
Expanded uncertainty in mm						0,043

The part's cylindricity deviation was measured by probing 11 points. The measurement values collected are shown in Table 11.

Table 11. Measurements of part cylindricity deviation using MMC

Measurements	Deviation Cylindricity (mm)
1	0,035
2	0,040
3	0,039
Average (mm)	0,038
Standard deviation (mm)	0,003

Table 12 gives the components of the measurement uncertainty for this measurand.

Table 12. Uncertainty of part cylindricity deviation using MMC

Uncertainty components						
Greatness	**Estimate**	**IT**	**DP**	**GL**	**CS**	**Standard uncertainty**
LMMC	**0,038**	A	**Normal**	2	1	**0,00153**
RMMC	**0,001**	B	**Rectangular**	O	1	**0,00029**
EA	**0,0029**	B	**Triangular**	O	1	**0,00119**
ICMMC	**0,00191**	B	**Normal**	11,744	1	**0,00191**
Combined standard uncertainty (u_c) in mm						0,00273
Effective degree of freedom veff						14,462
Scope factor k						2,145
Expanded uncertainty in mm						0,006

The measurement result is (0.038 ± 0.006) mm, for k = 2.14 and 95.45 % coverage.

The angles were calculated using the machine's programme, based on two lines. These were obtained by probing 5 points on two phases of the part. The measurement result is shown in Table 13.

Table 13. Angle measurements of the part using the MMC

Measurements	Angle
1	124° 39' 37"
2	124°43'50"
3	125° 39' 58"
Average	125° 01' 08"
Standard deviation	00° 33' 42"

Table 14 shows the evaluation of the uncertainty of the angle measurement in the MMC.

Table 14. Angle uncertainty data in the MMC

Uncertainty components						
Greatness	Estimate	IT	DP	GL	CS	Standard uncertainty (°)
LMMC	125° 01' 08"	A	Normal	2	1	0° 19' 27"
RMMC	0° 00' 01"	B	Rectangular	w	1	0° 00' 00"
Combined standard uncertainty (u_c) in °						0° 19' 27"
Effective degree of freedom veff						2,0000
Scope factor k						4,303
Expanded uncertainty in °						1° 23' 42"

Table 14 shows that the result of the angle measurement is (125° 01' 08" ± 1° 23' 42") for k = 4.30 and 95 % coverage.

4.2. Measurement with a profile projector

The profile projector (Fig. 7) used to take the measurements is from the manufacturer Carl Zeiss and has the following characteristics: model MP320; nominal range of 25 mm for linear measurements and 360° for angle measurements; resolution of 5 iim for linear measurements and 2 ' for angular measurements.

Figure 7. Profile projector

Figure 8 shows the measurands measured with the analysed part, which were a diameter and an angle.

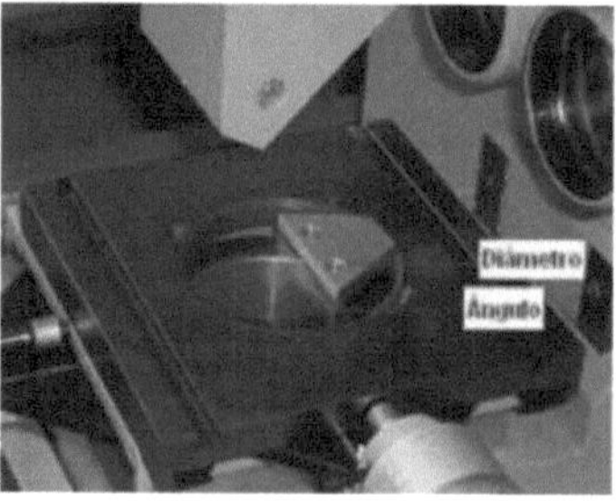

Figure 8. Angle and diameter measurement with profile projector

The values of the part diameter measurements are shown in Table 15.

Table 15. Part diameter measurements using the profile projector

Measurements	Readings (mm)	Diameters obtained (mm)
1	13,375	6,515
2	19,890	6,505
3	13,385	6,515
4	19,900	-
Average (mm)		6,512
Standard deviation (mm)		0,008

Table 16 gives the components of the measurement uncertainty for this measurand.

Table 16. Evaluation of part diameter uncertainty using the profile projector

Uncertainty components						
Greatness	Estimate	IT	DF	GL	CS	Standard uncertainty
LFr	6,512	A	Normal	2	1	0,00333
RFr	0,005	B	Rectangular	w	1	0,00289
"Fr	0,00001	B	Rectangular	w	3,907	6,35085E-08
"Fe	0,00002	B	Rectangular	w	3,907	1,38564E-07
AT	0,6	A	Normal	1E+10	0,000228	0,08660
Combined standard uncertainty (uc) in mm						0,0044
Effective degree of freedom veff						6,1247
Scope factor k						2,447
Expanded uncertainty in mm						0,011

The result of the diameter measurement with the profile projector is (6.512 ± 0.011) mm, for k = 2.48 and 95.45 % probability of coverage. The variable that most influenced the final uncertainty was the variability of the readings, because in this measurement system the operator contributes a significant portion of the errors.

The values of the part angle measurements are shown in Tab. 17.

Table 17. Part angle measurements using the profile projector

Measurements	Readings	Angles
1	272° 18'	67° 18' 00"
2	205 ° 00'	
3	272 ° 02'	67° 02' 00"
4	205 ° 00'	
5	272 ° 06'	67° 00' 00"
6	205° 06 '	
Average		67° 07'
Standard deviation		0° 10'

Table 18 gives the components of the measurement uncertainty for this measurand.

Table 18. Evaluation of part angle uncertainty using the profile projector

Uncertainty components						
Greatness	Estimate	IT	DP	GL	CS	Standard uncertainty
LPr	67° 07'	A	Normal	2	1	0° 05' 42"
RPr	00° 02'	B	Rectangular	w	1	0° 01' 09"
Combined standard uncertainty (u_c) in °						0 ° 05' 49"
Effective degree of freedom veff						2,1678
Scope factor k						4,303
Expanded uncertainty in °						0° 24' 00"

The expanded uncertainty associated with measuring the angle with the profile projector is 24' 0" for k = 4.30. The variable that most influenced the final uncertainty was the variability of the readings.

4.3. Measuring with a tooling microscope

The tooling microscope used to measure the characteristics of the screw (Fig. 9) is made by Carl Zeiss and has the following characteristics: a nominal range of 25 mm and a resolution of 1 pm for linear measurements. For angle measurements, the resolution is 1' and the nominal range is 360°.

Figure 9. Measurements of screw characteristics with a tooling microscope

The values of the screw thread fillet angle measurements are shown in Tab. 19.

Table 19. Measurements of the screw thread fillet angle using the tooling microscope

Measurements	Angle
1	59° 58' 00"
2	59° 50' 00"
3	59° 55' 00"
Average	59° 54'
Standard deviation	0° 04'

Table 20 gives the components of the measurement uncertainty for this measurand.

The result of the angle measurement with the tooling microscope is (59° 54' ± 0° 10'), for k = 4.30 and 95.45 % coverage. In this case, the variability of the readings contributed significantly to the final uncertainty.

Table 20. Evaluation of angle uncertainty using the tooling microscope

Uncertainty components						
Greatness	Estimate	IT	DP	GL	CS	Standard uncertainty
LMi	59° 54'	A	Normal	2	1	0° 02' 20"
RMi	01'	B	Rectangular	w	1	0° 00' 35"
Combined standard uncertainty (u_c) in °						0° 02' 24"
Effective degree of freedom veff						2,2524
Scope factor k						4,303
Expanded uncertainty in °						0° 10'

The values of the measurements of the external diameter of the screw are shown in Tab. 21.

Table 21. Screw outer diameter values using the tooling microscope

Measurements	Diameter (mm)
1	9,452
2	9,445
3	9,45
Average	9,449
Standard deviation	0,004

Table 22 gives the components of the measurement uncertainty for this measurand. It can be seen that, in this case, the expanded uncertainty is 0.009 mm for k = 4.30.

In this case too, the variability of the readings contributed more to the final uncertainty.

Table 22. Evaluation of the uncertainty of the external diameter of the part using the tooling microscope

Uncertainty components						
Greatness	Estimate	IT	DP	GL	CS	Standard uncertainty
LMi	9,449	A	Normal	2	1	0,00208
RMi	0,001	B	Rectangular	w	1	0,00057
Combined standard uncertainty (u_c) in °						0,00216
Effective degree of freedom veff						2,3195
Scope factor k						4,303
Expanded uncertainty in °						0,009

CHAPTER V- CONCLUSIONS

The following conclusions can be drawn:

The electronic spreadsheets developed implement the GUM measurement uncertainty assessment script, thus simplifying its calculation. They are also a reliable tool, as they reduce possible calculation errors that could occur if they were carried out manually.

As expected, *Excel* proved to be a powerful tool for implementing the calculation scripts, making it possible to obtain the final measurement uncertainty in an automated way.

Interpreting the GUM (2003) is not a simple task, since some of the information in this document is neither explicit nor easy to understand. As a result, assessing the uncertainty of the measurands with the measurement system under consideration becomes complex. In addition, surveying all the sources of influence on the measurement result of these measurands required an in-depth study of the measurement system used.

The spreadsheets are open, so users can modify them and adapt them to new applications.

Due to the ease of estimating uncertainty, provided by the spreadsheets, it is hoped to popularise the GUM and the script for calculating measurement uncertainty.

CHAPTER VI - BIBLIOGRAPHICAL REFERENCES

ABNT NBR ISO 9001 Quality management systems - Requirements. 2000.

ASSOCIAÇÃO BRASILEIRA DE NORMAS TÉCNICAS, NBR NM-ISO 1: **Standard reference temperature for industrial length measurements**, 1997. 2p.

ASSOCIAÇÃO BRASILEIRA DE NORMAS TÉCNICAS, NBR NM 216: **Pachymeters and depth gauges - Construction characteristics and metrological requirements**, 2000. 15p.

BARP, A. M. **Methodology for evaluating and managing the uncertainty of temperature measurement systems.** 2000. 126 p. Dissertation - Federal University of Santa Catarina, Florianópolis.

FRANCO, S. M. **Requirements for the application and evaluation of metrological reliability in metrology laboratories**. 1996. 110 p. Master's thesis - State University of Campinas, São Paulo.

GRACHANEN, C. **Uncertainty Calculator**. 2002. Available at: <http://metrologyforum.tm.agilent.com/download3.shtml>. Accessed on: 2 October 2010.

INMETRO. **International System of Units** - SI January 2007. 116p.

INMETRO, **Guia para Expressão da Incerteza de Medição,** Rio de Janeiro, 3ª Brazilian Edition, 2003. 120p.

INMETRO, **International Vocabulary of Metrology Fundamental and General Concepts and Associated Terms - VIM**, 2009. 77p.

JORNADA, D. H. **Implementation of a guide to measurement uncertainty for laboratory assessors in the RS metrological network.** 2009. 100 p. Master's thesis - Federal University of Rio Grande do Sul, Porto Alegre.

KACKER, R.; SOMMER, K.; KESSEL, R. Evolution of modern approaches to express uncertainty in measurement. **Metrologia**, Sevres, v. 44, p. 513-529, 2007.

KUNZMANN, H; WÄLDELE, F; NI, J. Accuracy Enhancement, in: BOSCH,

J.A.: Coordinate Measuring Machines and Systems, New York, Marcel Dekker, Inc., 1995

LEMOS, L.L., MORAES, M.A.F., SOUZA, C.C, VALDÉS, R.A. Mathematical modelling of measurement processes using calipers. XI Mechanical and Mechatronic Engineering Week. November 2009. Uberlândia.

LINK, W. Mechanical metrology: Expression of measurement uncertainty. 1997. p.174.

MATUSEVICH, A. INcerTI: A tool for calculating parameters and uncertainties in tensile testing. 2001. Available at: <http://www.inti.gob.ar/cordoba/boletin/boletin05/pdf/1 -3.pdf>. Accessed on: 5 October 2010.

MORAES, M.A.F., SOUZA, C.C., LEMOS, L.L., VALDÉS, R.A. Calibration of a dial indicator with a universal machine. XVII National Congress of Mechanical Engineering. CREEM2010. Viçosa.

MORAES, M.A.F., VALDÉS, A.R., LACERDA, H.B. Estimation of roughness measurement uncertainty and analysis of the influence of vibrations. Paper accepted for publication in **TC4 IMEKO** XVIII Symposium. 27 - 30 September 2011. Natal - RN. Brazil.

MORAES, M. A. F. Development of Electronic Spreadsheets to Calculate Measurement Uncertainty. 2011. 101 f. Monograph, Federal University of Uberlândia, Uberlândia.

NBR ISO/IEC 17025 **"General requirements for the competence of testing and calibration laboratories"**. January 2005.

OLIVEIRA, J. E. F.; MESQUITA, N.G.M. Development of a computer programme to control dimensional measurements with an emphasis on guaranteeing product conformity. **Innovare Magazine.** Campos Gerais, v. 8, Jul-Dec. 2009. Available at: <http://www.cescage.edu.br/new/main.php?module=edit edicoes>. Accessed on: 1 October 2010.

SALLUM, A. T. **Automated system for calibrating multifunctional measuring instruments.** 2002. 86 p. Dissertation - Federal University of Minas

Gerais, Minas Gerais.

STEMPNIAK, C. R. **Software for modelling, automation and evaluation of measurement systems.** 2004. 100 p. Master's thesis - Federal Technological University of Paraná, Paraná.

SOUZA, C.C, VALDÉS, A.R., COSTA, H.L. and PIRATELLI FILHO, A. **A contribution to the measurement of cylindricity and circularity deviation.** Paper accepted for publication at COBEM2011. October, 24-28. Natal - RN, Brazil.

VALDÉS R.A. e RIBEIRO J.R.S. Incerteza na Medição da Largura de Cordões de Solda. **Revista Soldagem Inspeção.** São Paulo, Vol. 14, No. 3, p.263-269, Jul/Sep 2009.

VIEIRA SATO, D. P. **Determining measurement uncertainty at three coordinates.** São Carlos School of Engineering, University of São Paulo, 110 p, São Paulo, Brazil, 2003.

I want morebooks!

Buy your books fast and straightforward online - at one of world's fastest growing online book stores! Environmentally sound due to Print-on-Demand technologies.

Buy your books online at
www.morebooks.shop

Kaufen Sie Ihre Bücher schnell und unkompliziert online – auf einer der am schnellsten wachsenden Buchhandelsplattformen weltweit! Dank Print-On-Demand umwelt- und ressourcenschonend produziert.

Bücher schneller online kaufen
www.morebooks.shop

Printed by Books on Demand GmbH, Norderstedt / Germany